NATIONAL GEOGRAPHIC

School Publishing

Saturn
The Ring World

PATHFINDER EDITION

By Lesley J. MacDonald

CONTENTS

Ring Wo

rld

● By Lesley J. MacDonald ●

Saturn is one of the best-known planets. Yet it is also one of the most mysterious. Now a spacecraft looks at this amazing ring world.

Have you ever wondered what it would be like to soar through Saturn's rings? A robot **spacecraft** got the chance to do that. It is named Cassini.

In 2004, Cassini arrived near Saturn. The spacecraft had spent more than seven years speeding through space. It had traveled more than 3.5 million kilometers (2.2 million miles). Now it faced the most dangerous part of its mission.

Cassini arrived below Saturn's rings. It had to fly through them. The rings are made of millions of chunks of ice and ice-covered rock. Some pieces are the size of a grain of sand. But others are as big as a house. Could Cassini avoid a collision?

Luckily, Cassini made it through the rings without a scratch. The spacecraft then started to **orbit**, or go around, Saturn.

Cassini isn't the first spacecraft to go to Saturn. Voyager 2 visited the ring world in 1981. But it didn't hang around for long. After snapping hundreds of photos, it zoomed away.

Cassini, on the other hand, isn't going anywhere else. It completed its mission to orbit Saturn for four years. Now it is on an extended mision. There's a lot scientists want to learn about Saturn and its rings!

A Gas Giant

Saturn is the second largest **planet** in our solar system. A planet is a large object that orbits a star. Only Jupiter is bigger. Saturn is so large that 750 Earth-size planets could be placed inside it.

Like its larger neighbor, Saturn is made of gases. All we see on Saturn are clouds of gas. However, a small solid core may be buried deep below those cloud tops.

The gases in Saturn's **atmosphere** may look calm. They aren't. They swirl around the sky. Clouds, tornadoes, and hurricanes whip around Saturn. At the equator, winds can reach around 1,600 kilometers (1,000 miles) an hour.

Lord of the Rings

Saturn is best known for its beautiful rings. It is not the only planet that has rings. Three other planets—Jupiter, Uranus, and Neptune— also have rings. However, those rings are much thinner and harder to spot.

From far away, Saturn's rings look almost like a single, perfect bracelet. But up close, things are much different. The planet actually has about a thousand different rings. Some of them are even braided together.

Tiny **moonlets** orbit alongside some of the rings. Their gravity may help keep the rings together. Without the moonlets, particles that make up the rings might float away.

Boldly Going. The Cassini spacecraft rockets toward Saturn.

Ring Around the Planet.
Saturn's famous ring is really about a thousand different bands. These colors show some of them.

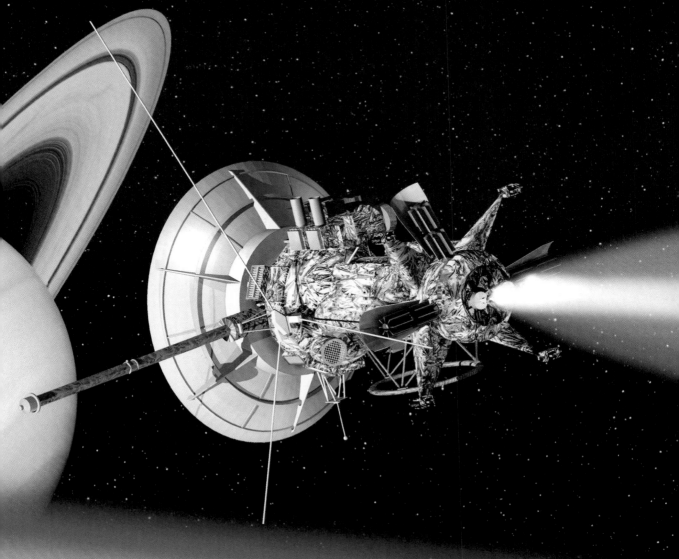

Enceladus

Titan

Many Moons. Voyager 2 flew by Saturn in 1981. Its images appear in this NASA collage. Now Cassini is taking new, clearer photos.

Rhea

Saturn

Dione

Mimas

Tethys

Meet the Moons

Saturn has more than 50 **moons**. Cassini discovered two of them. And the number of moons may continue to grow.

Scientists still have much to learn about Saturn's moons. Most of them are pretty small. Not many are more than 966 kilometers (600 miles) across. Earth's moon, in comparison, is 4,023 kilometers (2,500 miles) across.

Cassini is giving us clearer pictures of Saturn's moons. What are the moons like?

Enceladus is one of the shiniest objects in our solar system. Its icy surface reflects lots of light from the sun.

Mimas has a huge crater. It's 129 kilometers (80 miles) wide and 9.6 kilometers (6 miles) deep. The crater covers nearly a third of the moon's surface.

Tethys has long, deep trenches. It also has tiny moonlets. Imagine, a moon with moons!

Moon of Mystery

The most amazing moon is Titan. It is the second largest moon in our solar system. And it may be the most fantastic. A thick atmosphere covers Titan. The air is mostly nitrogen gas.

Some scientists think that life may exist on Titan's surface. To find out, Cassini is giving the moon extra attention. The spacecraft looked closely at Titan. Cassini gave us our first glimpse of the moon's surface.

Cassini saw dark streaks over part of the giant moon. What caused them? The answer may be blowing in the wind. Perhaps strong winds moved material across the surface.

Cassini also saw large dark blotches on Titan. They might be lakes. If so, they're not full of water. Titan is too cold for liquid water. Instead the lakes may be filled with liquid methane. On Earth, methane is normally a gas.

Safe Landing. This picture shows the Huygens spacecraft landing on Titan.

The Mission Continues

Scientists still want to learn more about Titan. The pictures that Cassini snapped do not show many details. One lake looks like a large cat head. Other features look either light or dark. Scientists wanted a closer look at the surface.

That's why Cassini brought a tiny spacecraft to Saturn. It is named Huygens. The ship was made to land on Titan.

After a bumpy ride, Huygens reached the surface of Titan. Then it began gathering information—lots of it. The spacecraft studied both the surface and the atmosphere. It found many rivers on the moon's surface.

Cassini has just begun its great adventure. We don't know what other amazing discoveries it will make. Perhaps one day, people will follow the spacecraft to Saturn. We can only imagine the sights they will see.

Wordwise

atmosphere: layer of air around a planet or moon

moon: large object that goes around a planet

moonlet: tiny moon

orbit: to go around

planet: large object that goes around a star

spacecraft: ship that flies through space

What's in a

The Cassini and Huygens spacecraft give us amazing views into space. They show everything from Saturn's storms to moonlets dodging around the planet. But what's the story behind these spacecraft? Why do they have such odd names?

The spacecraft are named after scientists who changed the way people think about Saturn. Christiaan Huygens and Gian Domenico Cassini lived a long time ago. Yet their discoveries about this ringed planet were truly out of this world.

Ring Research. Huygens BELOW discovered Saturn's rings. This drawing by Huygens shows Saturn orbiting around the sun RIGHT.

Finding Rings

Huygens and Cassini lived in the 1600s. At the time, people did not know much about space. Scientists didn't have spacecraft or cameras. Simple telescopes were the tools of the trade.

These were just tubes with pieces of curved glass inside. Yet in the right hands, they could show amazing details of space.

Huygens used such a telescope. With it, he saw things that no one had seen before. In 1655, he discovered Titan, Saturn's largest moon. The very next year, Huygens figured out that rings surrounded Saturn.

At first, people thought he was crazy. Who had ever heard of rings around a planet? Yet before long, his idea caught on. Saturn soon became famous for its rings.

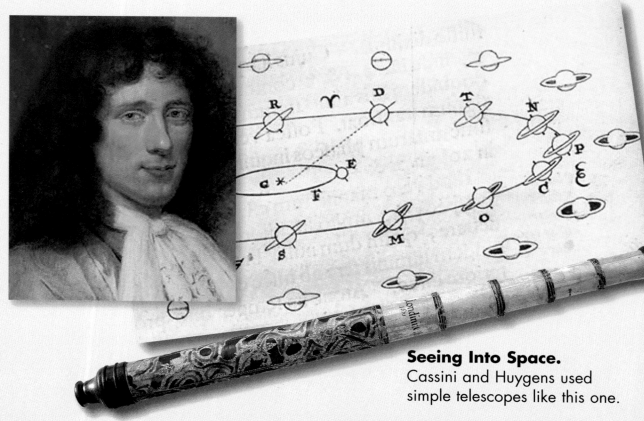

Seeing Into Space. Cassini and Huygens used simple telescopes like this one.

Name?

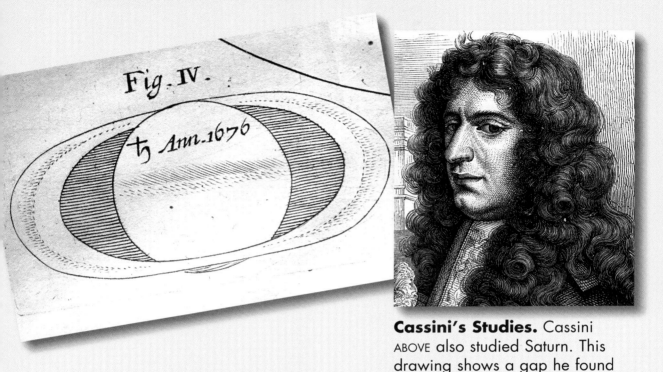

Fig. IV.

♄ Ann. 1676

Cassini's Studies. Cassini ABOVE also studied Saturn. This drawing shows a gap he found in the rings LEFT.

Filling the Gap

Gian Domenico Cassini lived at about the same time as Huygens. During the mid-1600s, both men worked in Paris, France. They both studied the stars with a huge telescope there.

Cassini first made a name for himself with discoveries about Jupiter and Mars. Yet Cassini may be best remembered for his studies of Saturn. He figured out that Saturn's rings were made of rocky material. He discovered four of Saturn's moons. He was also the first to see a gap in Saturn's ring system.

Today, this big space between the rings is called the Cassini Division. Cassini's amazing discoveries made him one of the most respected scientists of his day.

Days of Discovery

Cassini and Huygens used the best telescopes available. But they also used their brains. First these scientists studied the sky. Then they put their math skills to work. They did calculations to figure out how the planets moved.

By today's standards, Cassini and Huygens used simple tools. Yet they were able to figure out how planets looked—and how they moved. Together, these men changed the way people understood space.

Scientists like Cassini and Huygens made amazing discoveries. But there is still much to learn. Scientists find out more each day. Yet every discovery leads to new questions. Someday you might help to find the answers.

Changing How

A Mountaintop View.
This building protects
a powerful telescope
on a mountaintop
in Hawaii.

M uch has changed since the time when Cassini and Huygens studied space. Today we have tools that let us study planets and stars in ways these scientists could never have imagined.

Not only do we have powerful telescopes on Earth. We also have telescopes scooting around in space. We even have spacecraft with cameras that send us pictures of worlds beyond Earth.

Tools like these let us see space like never before. Each gives us new information about our universe.

Views From Earth

The telescopes we use on Earth come in all sizes. Small telescopes let people study planets and stars from their own backyards. But serious scientists need telescopes with more power.

Some of these supersize telescopes are on mountaintops. There the sky is darker and clearer than in cities. Yet the sky can make distant things look fuzzy—and some things are just too far away to be seen clearly. So scientists have found ways to get a better view. They send telescopes into space.